Virginia Flowers

by

Tracey Y Hughes

ISBN: 0983427666
ISBN-13: 978-0-9834276-6-7

***

www.deepseapublishing.com

***

Printed in the United States of America

Dedication

This book is dedicated to my loving parents. Not only did they give me a wonderful childhood, they also gave me the beautiful landscapes and gardens from which this work was derived. However, this dedication would not be complete without recognizing the original artist, God, our Creator, our Lord and Savior. In applying, 1 Corinthians 3:7; I nor any other photographer or artist creates these beautiful masterpieces. We can only use our talents in an attempt to capture their beauty and pass them on to you in a chosen medium. I pray that everyone takes the time to look at the magnificent masterpieces created for all of us.

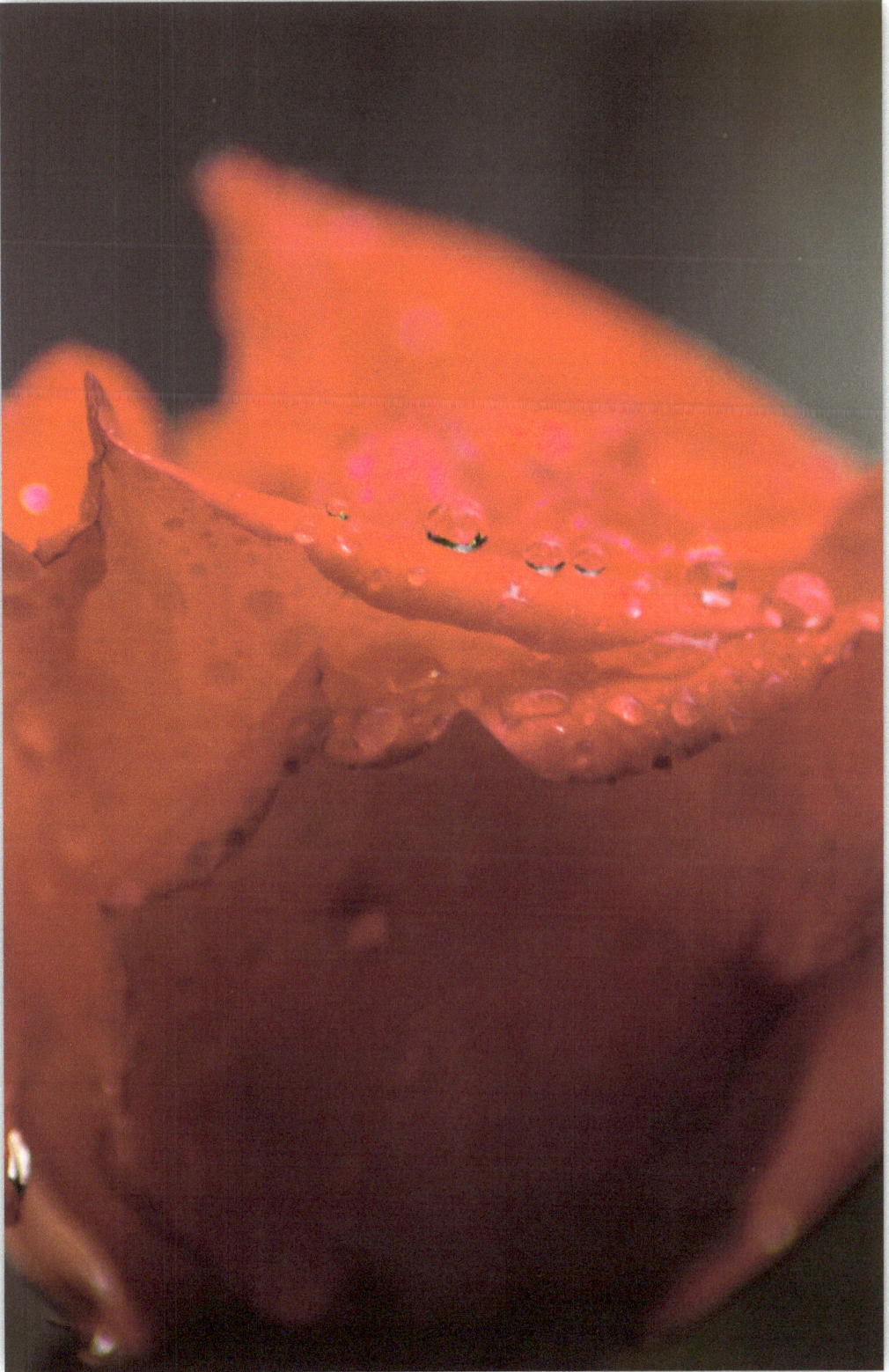

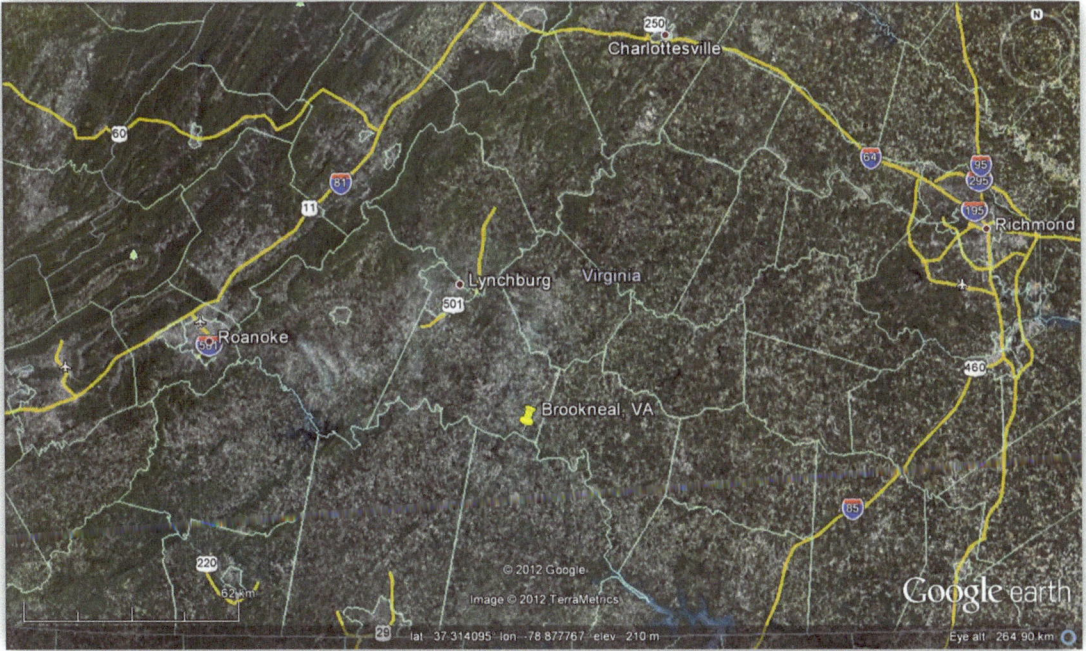

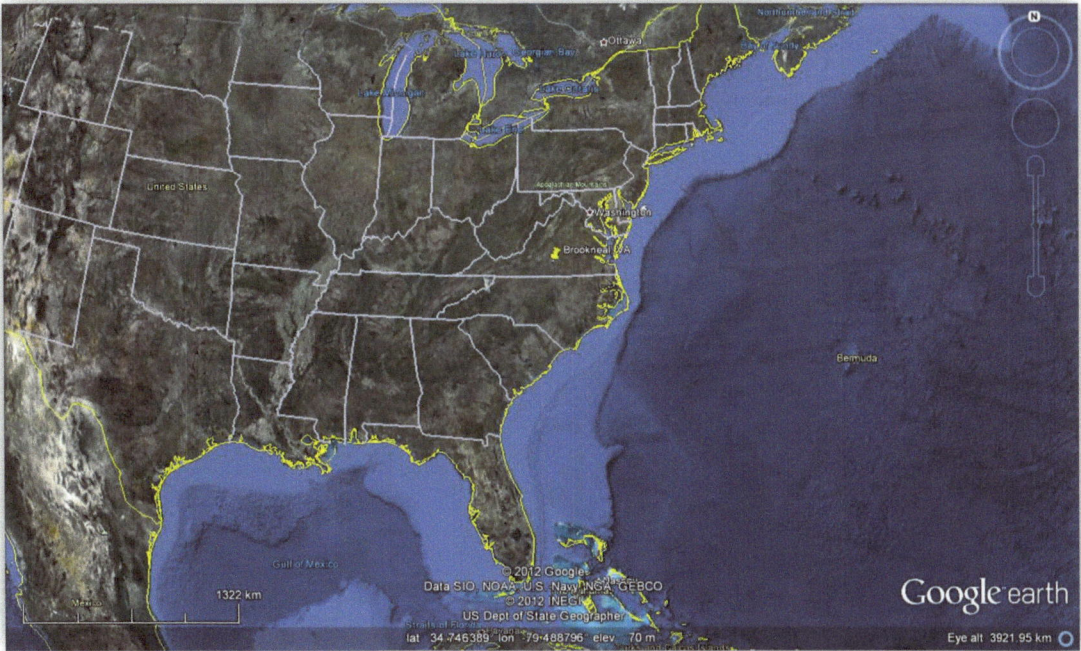

The above maps are courtesy of Google Earth.

All flowers in these pictures were taken in Campbell County, Virginia.   Most were taken at my parent's home near Brookneal, VA. I am not a botanist, flower grower, or have any expertise when it comes to flowers. But I do love to take pictures of flowers from a perspective that is often unseen by many people. I hope that you enjoy this book.

I have attempted to provide the correct name of each flower in the table below.

Flower	Page	Flower Name
	Cover	Bachelor Buttons
	Title Page	Lantana
	4	Eastern Purple Coneflower (Echinacea purpurea)
	5	Flax

Flower	Page	Flower Name
	6	Pansy
	7	Gerbara Daisy
	7	Gerbara Daisy
	8	Sunflower
	9	Cardinal Climber
	9	Eastern Purple Coneflower

Flower	Page	Flower Name
	10	Rose
	11	Clematis
	12	Bachelor buttons (Cornflower)
	13	Clematis
	13	Impatiens

Flower	Page	Flower Name
	14	Clematis
	15	Lily
	16	Sweet Williams
	17	Bachelor buttons (Cornflower)

Flower	Page	Flower Name
	18	Day Lily
	19	Clematis
	19	Rose
	20	Lilies in the foreground
	21	Royal Purple Zinnia

Flower	Page	Flower Name
	22	Iris
	23	Rose
	23	Rose
	24	Eastern Purple Coneflower
	25	Butterfly Bush

Flower	Page	Flower Name
	26	Rose
	26	Black-Eyed Susan
	27	Rose
	28	Eastern Purple Coneflower
	29	Eastern Purple Coneflower
	29	Eastern Purple Coneflower

Flower	Page	Flower Name
	30	Iris
	31	Crown Vetch
	32	Clematis
	32	Spider on a sunflower
	33	Red Hot Poker

Flower	Page	Flower Name
	34	Rose
	35	Iris
	36	Iris
	37	Aster
	37	Day Lily

Flower	Page	Flower Name
	38	Coneflower

About the Author/Photographer

Tracey Yvette Hughes was born in Halifax, Virginia and was raised in Campbell County, Virginia. She is a devoted mother of two smart and pretty teenage daughters. Tracey attended college at University of California, Northridge and high school at William Campbell High School. She loves the outdoors and enjoys capturing those shots of rare moments in nature. Tracey now lives in Lynchburg, Virginia.

From an early age, Tracey started snapping photos of everything. As a young adult, she began to focus more on quality shots of landscape, flowers, wildlife and portraits in natural settings. This is Tracey's first book of flowers.

To see more of her works, check out her author's page on www.deepseapublishing.com. She also had a photography book with many of the animals and insects found in Virginia. The book, titled ***Virginia Wildlife***, is also available at Deep Sea's online store and Amazon.com.

Deep Sea Publishing

Deep Sea Publishing, LLC is a Florida company with offices in Florida and Virginia. The company publishes:

- Fictional Novels,
- Historical Fiction,
- Children's Books,
- Young Adult and Teen Fiction,
- Technical References,
- Photography Books,
- and more.

Deep Sea Publishing is a USA firm owned by 100% USA citizens. All printing and production is in the USA. Other works from Deep Sea Publishing include:

- *The Bryant Family Chronicles: Death and Gold in Zara Zote*
- *The Gallivan Legacy*
- *Hardt's Tale*
- *Let Sleeping Dragons Lie*
- *The Good Fight*
- *Seven Summits: The Magical Talent*
- *Not Myself*
- *Carcharhinus obscurus, & Carcharias Taurus (kids' shark books)*
- *My Seasons with Grandma Francesca and Grandma Louise (children's book)*
- *Christmas in Paradise (children's book).*

To find out more about Deep Sea Publishing or to order books, visit our website at www.deepseapublishing.com.

www.ingramcontent.com/pod-product-compliance
Lightning Source LLC
Chambersburg PA
CBHW060831270326
41933CB00002B/50